Zero the Hero Saves the Day

Written by Denise K. Savidge
Illustrations by Deanna McRae
Printed using CreateSpace

Printed in the U.S.A.

This edition first printing, September 2015

The text for this book is set in Bookman Old Style.

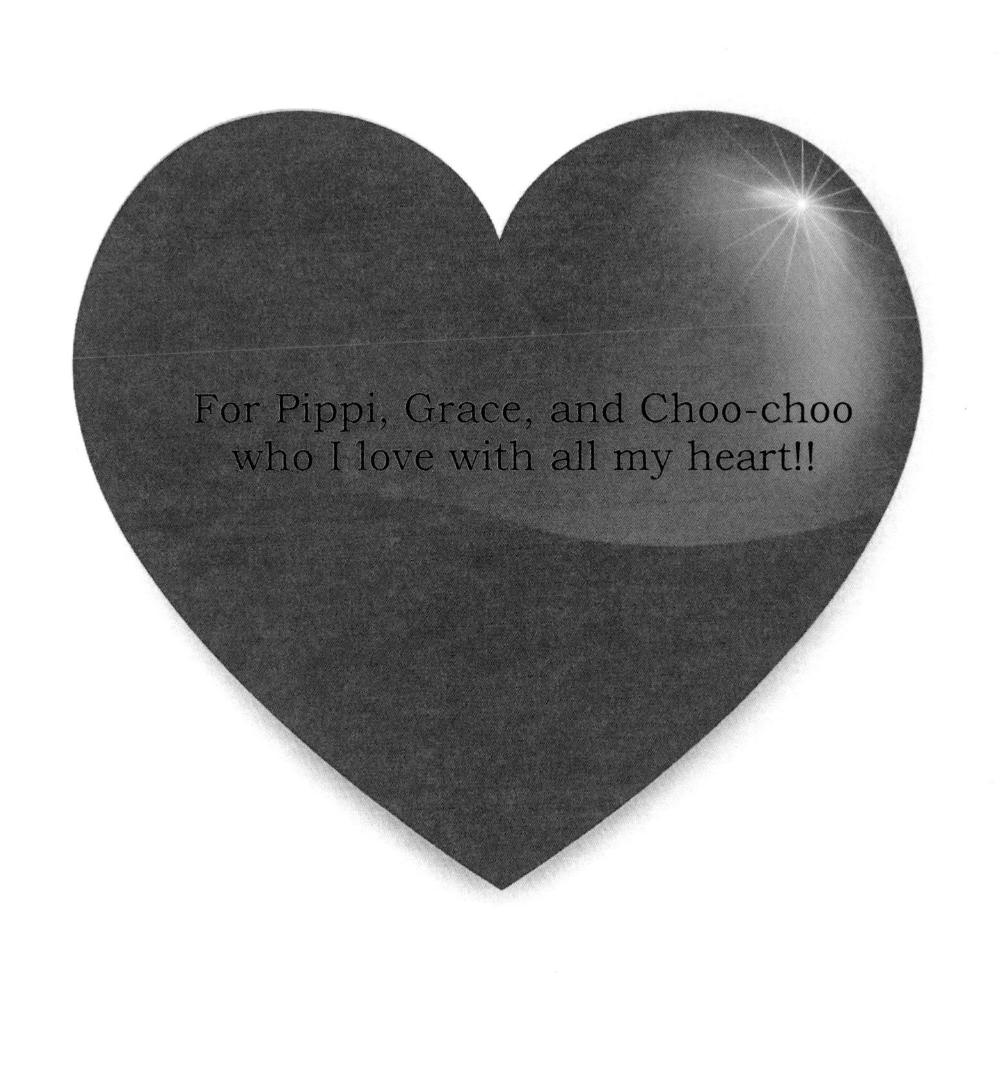
For Pippi, Grace, and Choo-choo
who I love with all my heart!!

The numbers were worried and all in a tizzy.
Seven was thinking so hard he was dizzy.
Up until now counting days was just fine,
Beginning with "One" and ending at "Nine."

But that day,
yes that day, the
troubles began,
On the tenth
day of school
with all
numerals on
hand.
Yet not one of
them knew how
to number this
day
Was it Nine's
job, or Eight's?
Maybe Two's?
Oy vey!

Oh the shock, oh the horror,
how could this be the case?
Every number was sure of its
number line place.
“All along we have been quite
enough to show sets,
We’ve shown groups and
collections. We’re as good as
it gets!”

"How can there be a number that not one of us 'is'?
There is no one more expert than us in this biz.
We can label Eight whiskers on a kitty cat's face,
We can count Nine stars twinkling high up in space."

"Every face has Two eyes and Two ears and One nose.
Every hand has Five fingers and each foot has Five toes
Counting sheep is the way we help all kids to slumber.
Up to now, we could label each 'thing' with its number."

“We have never had trouble in our whole number history.
But this day we call ‘Ten’ is a baffling mystery.”
They all shook their heads with despair and great gloom.
“This tenth day spells chaos, we see nothing but doom.”

Tiny One to this
point had not
joined the
discussion.
Pulling thumb
from her
mouth, she
then threw this
thought in:
“Is it possible ...
likely ... could
their feasibly
be...
Any use for
that number
that’s one less
than me?”

"Less than One? That is nothing," all her friends began roaring.
"Who would ever choose to be something so boring?
Nothing is nil, it is zilch, it is naught.
It's a nobody! Why it can hardly be taught."

Little One stood her ground in the face of such scorn.
"I may only be One, in your eyes barely born.
He's my friend, that old fellow with the large circle face,
And he won't mind one bit if he only holds place."

“Standing for nothing is his claim to fame.
He’s NOT an old nobody! Zero’s his name.
Hanging out with me, little One, is his yen.
And both of us, paired up, you could call ‘Ten’.”

The numbers got silent, all mouths were agape.
They stood dumbfounded, astonished,
even old number Eight,
(Considered the wisest of all of the numbers)
Stood feeling as if he was dull, **dumb** and **dumber**.

Then, “One, you’re a genius. Go call on your friend.
Tell him we would be happy to make you both Ten.
He can stand with us all, though his worth will be nil
We can then count to twenty, and then higher still.”

"With Three he makes thirty, and then with a Four,
We can count on to forty, and fifty and more.
Next to me he'll be eighty, just holding his spot.
He'll be greater than nothing, no longer be 'naught'."

“He’ll be Zero our Hero, known far and known wide
As the number we’re honored to have by our side.
For with him we are greater than we ever could be
Counting onward and upward toward infinity!”

Tiny One (who was not one to see lots of action)
Smiled a smile of great satisfaction.
For her dear old friend Zero was now one of the group,
And SHE was the one who had gotten the scoop.

Old Zero, he glowed, just so happy and proud
Swollen with pride, he rolled through the crowd
He'd started the day as a nothing, a Zero.
But from then on was known as the
Number Line Hero!

The
End ...
or is it?

Denise K. Savidge, MEd

Denise K. Savidge has been an educator of children preschool to high school aged since 1992. Currently she is the Academic Dean of a boarding school for middle school boys in Westminster, South Carolina. She lives with her three children, four cats, seven koi and one beta fish and is working on her PhD at Clemson University. Not surprisingly, it took eight years to complete this book with her crazy life.

Watch for more Zero the Hero titles
coming soon...

Zero the Hero Gains Power
Zero the Hero Goes Berserk

Made in United States
North Haven, CT
17 June 2023